CONSERVATION CHAMPIONS!

CONSERVING WATER

SADIE SILVA

Enslow PUBLISHING

Please visit our website, www.enslow.com.
For a free color catalog of all our high-quality books, call toll free 1-800-398-2504 or fax 1-877-980-4454.

Library of Congress Cataloging-in-Publication Data
Names: Silva, Sadie author
Title: Conserving water / Sadie Silva.
Description: Buffalo, NY : Enslow Publishing, [2027] | Series: Conservation champions! | Includes index.
Identifiers: LCCN 2025053947 (print) | LCCN 2025053948 (ebook) | ISBN 9781978547742 library binding | ISBN 9781978547735 paperback | ISBN 9781978547759 ebook
Subjects: LCSH: Water conservation–Juvenile literature
Classification: LCC TD388 .S57 2027 (print) | LCC TD388 (ebook) | DDC 333.91/16–dc23/eng/20260203
LC record available at https://lccn.loc.gov/2025053947
LC ebook record available at https://lccn.loc.gov/2025053948

Published in 2027 by
Enslow Publishing
2544 Clinton Street
Buffalo, NY 14224

Designer: Tanya Dellaccio Keeney
Editor: Therese Shea

Photo credits: Cover (girl) AzmanL/iStock.com; cover (faucet) alma.tross/Shutterstock.com; p. 5 FamVeld/Shutterstock.com; p. 7 Prostock-studio/Shutterstock.com; p. 9 fizkes/Shutterstock.com; p. 11 (top) chaossart/Shutterstock.com; p. 11 (bottom) Bilanol/Shutterstock.com; p. 13 Parilov/Shutterstock.com; p. 15 dr.Barmely/Shutterstock.com; p. 17 (top) Yalcin Sonat/Shutterstock.com; p. 17 (bottom) Stefan_Sutka/Shutterstock.com; p. 19 VladKK/Shutterstock.com; p. 21 Creativa Images/Shutterstock.com.

Printed in the United States of America

CPSIA compliance information: Batch #CSENS27: For further information contact Enslow Publishing, at 1-800-398-2504.

CONTENTS

BOLDFACE WORDS APPEAR IN THE GLOSSARY.

WATER IS IMPORTANT

We use water for so much. We wash our hands and bodies with water. Toilets **flush** our body waste away with water. We wash our clothes and dishes with water. We give water to our plants and flowers. We need to drink water to keep our bodies healthy too.

WATER IS ALSO A SOURCE OF FUN!

CONSERVE IT

Do you ever think about where your water comes from? Not everyone has **access** to clean water. Not everyone can just turn on the tap. People who have access to water should conserve it. That means not waste it.

MOST PEOPLE IN THE UNITED STATES CAN ACCESS CLEAN, RUNNING WATER. ABOUT 2 MILLION AMERICANS DON'T HAVE CLEAN, RUNNING WATER, THOUGH.

A SMALL AMOUNT

Only about one percent of Earth's water is fresh water we can use. The rest is either frozen fresh water or salt water, like the salt water in the world's oceans. Even in places with lots of fresh water, water should be conserved. It saves **energy** and money.

EARTH HAS A LOT OF WATER, BUT ONLY A SMALL AMOUNT IS USABLE.

WHERE IT COMES FROM

Fresh water falls as rain and other forms of **precipitation**. It collects in lakes, ponds, rivers, streams, and under the ground. When it's needed, it's taken from a source to a water treatment plant. There, it goes through a **process** to become safe for people's use.

FRESH WATER IN NATURE MAY HAVE THINGS IN IT THAT CAN MAKE US SICK. IT NEEDS TO BE TREATED BEFORE WE USE IT.

WATER TREATMENT PLANT

GETTING IT READY

A lot of energy is needed to pump water where it needs to go. It also takes many **resources** to clean it, such as special **chemicals**. Wasting clean water takes it out of the usable supply. It then must go through the treatment plant before it's used again. This is a costly process.

AFTER YOU FLUSH THE TOILET OR RUN THE TAP, THE WATER GOES BACK TO A WATER TREATMENT PLANT WHERE IT'S CLEANED. OFTEN IT THEN GOES BACK INTO NATURE UNTIL IT'S NEEDED AGAIN.

WAYS TO CONSERVE

How can you conserve water? Use only what you need. Don't let the tap run while you brush your teeth. Take short showers instead of baths. Collect rainwater to care for plants instead of using a hose. Using less water will save your family money too!

RAIN BARRELS LIKE THIS ONE COLLECT RAIN THAT YOU CAN USE FOR YOUR PLANTS AND FLOWERS.

HIDDEN USES

A lot of water is used in people's homes, but **industries** use a lot of water too. The **agriculture** industry uses the most. The clothing industry uses the second most. Some are trying new methods to conserve water, such as planting crops that need less water.

FARMS IN PLACES WITH LESS RAIN MAY USE A LOT OF WATER FOR THEIR CROPS.

TALK ABOUT IT

Tell friends and family what you've learned about conserving water. Look up water facts that will make them want to be conservation heroes too. One person conserving water can help. Many people working together can make a bigger difference!

TAKING A 5-MINUTE SHOWER USES ABOUT 10 GALLONS (38 L) OF WATER. A SMALL BATHTUB USES ABOUT 40 GALLONS (150 L)!

ACTIVITY

- Decide on an idea for conserving water that you'd like to try.
- Make a sign asking others to try your idea.
- Add an interesting fact about water.
- For example, put a note above the bathroom sink reminding people to turn off the water while brushing their teeth. Tell them they can save 8 gallons (30 l) of water a day!

Dont forget
to save
water

GLOSSARY

access: Able to reach.

agriculture: The science or job of farming.

chemical: Matter that can be mixed with other matter to cause changes.

energy: Power, such as electricity, used to do work.

flush: To clean with water.

industry: A group of businesses that do a similar thing.

precipitation: Rain, snow, sleet, or hail.

process: A series of steps or actions taken to complete something.

resource: A usable supply of something.

source: The starting point of something.

FOR MORE INFORMATION

BOOKS

Amstutz, Lisa J. *10 Ways to Use Less Water.* North Mankato, MN: Pebble, 2024.

Gerry, Lisa. *Water! Why Every Drop Counts and How You Can Start Making Waves to Protect It.* Washington, DC: National Geographic Kids, 2023.

WEBSITES

Fresh Water Access
education.nationalgeographic.org/resource/freshwater-access/
Read more about the challenges of freshwater access.

Test Your Water Sense!
19january2017snapshot.epa.gov/www3/watersense/kids/games.html
Take a quiz about water conservation.

INDEX